Waterfalls and Rapids

First English language edition published in 1998 by
New Holland (Publishers) Ltd
London - Cape Town - Sydney - Singapore

24 Nutford Place
London W1H 6DQ
United Kingdom

80 McKenzie Street
Cape Town 8001
South Africa

3/2 Aquatic Drive
Frenchs Forest, NSW 2086
Australia

First published in 1997 in The Netherlands as
Watervallen en Stroomversnellingen by
Holding B. van Dobbenburgh bv, Nieuwkoop,
The Netherlands
Written by: Marianne C. Eelman
Translated from the Dutch by: K.M.M. Hudson-Brazenall

Copyright © 1997 in text: Holding B. van Dobbenburgh bv,
Nieuwkoop, The Netherlands

Copyright © 1997 in photographs: individual
photographers and/or their agents as listed on page 2

Copyright © 1997: Holding B. van Dobbenburgh bv,
Nieuwkoop, The Netherlands

All rights reserved. No part of this publication may be
reproduced, stored in a retrieval system or transmitted, in
any form or by any means, electronic, mechanical, photo-
copying, recording or otherwise, without the prior written
permission of the publishers and copyright holders.

ISBN 1-85368-696-4

Editorial direction: D-Books International Publishing
Design: Meijster Design bv
Cover design: M.T. van Dobbenburgh

Reproduction by Unifoto International Pty, Ltd

Technical Production by D-Books International
Publishing/Agora United Graphic Services bv

Printed and bound in Spain by Egedsa, Sabadell

CONTENTS

Introduction 3

1 The Origins of the Waterfall 4
Conditions needed for rapids and waterfalls 6
Where does all the water come from? 6

2 River Basins 13
The life cycle of a river 13
Papua New Guinea: a profusion of waterfalls 14
Exmoor, England: green country divided by
  fast-flowing rivers 15
The power of falling and flowing water 18
Rock formed by water: travertine 23
Geological structures that cause waterfalls 29
Collapsed limestone caves 34
Waterfalls that run dry 35

3 World Famous Waterfalls 38
Niagara Falls 38
Victoria Falls 42
The Iguaçu Falls 43
The Yellowstone River 46
Yosemite: snow and water flow over
  granite rocks 49
Natural barriers 52
China: a secret world behind the waterfall 53
The Angel Falls 53
Iceland: land of waterfalls 58

4 Flora and Fauna 63
Natural selection 63
Fauna in cold flowing water 63
Plants that defy falling water 68
The waterfall as a source of food 68
Waterfalls, rapids and man 69
Shooting the rapids 69
Switzerland: the Rhine Falls 78
Hydro-electricity 78

PHOTO CREDITS

American Express Travel, 40/41; Arkel, J. van/Foto Natura, 36[b], 37; Beekman, F., 39; Berg, A. van den/Foto Natura, 10[t], 36[b], 45, 52; Blankers, E./Foto Natura, 67; Corley, M., 54, 55[t]; Dennis, N.J./Foto Natura, 42[t]; Dobbenburgh, B. van, 13; Dobbenburgh, M.T. van, 7 10[l], 12, 21[b], 22, 47, 48, 60, 61, 62, 71[b], 74, 77; Eelman, M., 5, 29, 30, 31, 50[t], 55[b], 70; Ellinger, D./Foto Natura 75; Fey, T./Foto Natura, 71[t]; Hazelhoff, F.F./Foto Natura, 11, 15, 16[l], 21[t], 23, 24/25, 26, 27, 36[t] 59; Helo, P., 14, 20[b], 79; Jongman, P./Foto Natura, 19; Keereweer, A., 4, 28[t], 72/73; King, V., 43, 51, 56/57; Kluiters, J./Foto Natura, 18; 44; Meinderts, W.A.M./Foto Natura, 8/9, 16, 76[b]; P'lking, F./Foto Natura, 17; Schenk, A., 28[b], 68; Schwier, P.K., 53; Tromp, H./Foto Natura, 34; Vogelenzang, L./Foto Natura, 33, 64, 65, 66; Weenink, W./Foto Natura, 42[b]; Wegner, J./Foto Natura, 20[t], 35;

r=right, l=left, t=top, b=bottom, c=centre

Introduction

Waterfalls... thundering magic wreathed in white mist.

Water, who can survive without it? 71% of our planet consists of oceans filled with salt water; humans consist of some 65% water. It can be found all around us in an infinite number of states: as waves and tides, which gradually shape the coastlines of the earth; as ice capable of breaking massive rocks; as glaciers and rivers that force their way through mountains; as rain that falls from the sky in merciless downpours and waterfalls that plunge thunderously into the depths. Many of these phenomena associated with water affect people in their daily lives and arouse their imagination. The waterfall is one phenomena that arouses the imagination the most. A thunderous roar accompanies the falling water, splashing water droplets that wreath the area in white mists, rays of sunlight turning to rainbows by the reflection of the water, crumbling rocks and impressive views. The Swiss botanist, Robert Chodat (1865–1934) described the awesome grandeur of the Iguaçu waterfalls in Brazil as follows, 'If we stand at the foot of this world of cascades and lift our gaze ninety metres higher to the horizon which is nothing but water, then this awe-inspiring spectacle of an ocean plunging into the depths is almost frightening.' Charles Dickens, the English novelist (1812–1870) wrote after his visit to Niagara Falls, 'What voices spoke from under this roaring water; which faces, obscured on Earth, watched me from the glistening depths; what heavenly promise glittered in those angel's tears, the multi-coloured droplets that flew about and grasped the beautiful ever-changing rainbows.'

This book is about waterfalls and rapids, how they occur, where all the tumbling water comes from and how they disappear.

1 The Origins of the Waterfall

Without water there would be no waterfalls or rapids. Waterfalls and rapids are but a small part of a greater entity: the river. A river is a mass of water that flows to a lower level via a naturally formed course. As water always flows to the lowest point, most rivers begin in the mountains and finally empty into the sea.

Rivers continuously change their form according to the amount of water that they carry during wet and dry periods and to the amount of sediment they transport. They are, therefore, dynamic elements in the landscape. Cutting a cross-section through a river bed will, at any point on the river, provide an impression of the prevailing current. Small streams, like mountain streams, are often as deep as they are wide, whilst a river further downstream is often many times wider than it is deep. By measuring the vertical distance between two points in its course, we can obtain a measure of the hydraulic gradient (or slope) of the river bed, which is expressed in metres per kilometres. The further downstream one goes, the more the average gradient of a river bed smoothes out, whilst close to the source the gradient is steeper. The average gradient of a steep mountain stream can be as much as 60 metres per kilometres or more, whilst the gradient at the river mouth can be 0.1 metre per kilometre or less. It logically follows then that most waterfalls are found in the mountains. The longitudinal profile of

Hissing, roaring, misty and magical... these words are often used to describe the Brandy Wine Falls in Canada. Water has carved its way into solidified lava flows for tens of thousands of years to create these waterfalls. In the process a large hole, called a plunge pool, has formed at the foot of the falls. The falls are fed by meltwater from the glacier higher up, with most of the water spilling down in the spring, at a rate of more than 8,000 litres per second.

a developed river (a line that reflects the river surface from the source to the mouth of the river) is a curve with a decreasing gradient downstream. However, it is not a smooth, gradual curve because there are always local variations in the way a river drops. Variations in the longitudinal profile of a river can be the result of falls in sea levels which increase the gradient downstream. The reason may also lie in differences in resistance to erosion of the underlying rocks through which the river is trying to cut. Variations can also be the result of vertical movements along a fault line that cut diagonally across the river, or land movements which deposit material in the river and create a temporary dam. The most conspicuous variations in river beds, are carved out of solid rock, and cause the water to flow rapidly and turbulently through rapids, or cause water to fall over a steep precipice; thus the waterfall is born.

Yosemite National Park in California, USA, is home to many waterfalls that are fed with meltwater, particulary in the spring. In the foreground are the Vernal Falls with Nevada Falls lying behind. Both are formed by the Merced River flowing over a glacially scoured staircase. In the distance stand the snow-covered peaks of the Sierra Nevada.

▷ *Vallée du Trient, France. In the spring, many waterfalls appear in the mountains because of the enormous amount of meltwater finding its way down to lower-lying regions. Most of these falls dry up again in the summer.*

Conditions needed for rapids and waterfalls

To create rapids or a waterfall, a number of criteria must be met. There has to be a variation in the gradient of the river, which basically means that there has to be an abrupt change in the topography of the river course. A waterfall requires a steep precipice and rapids need obstacles in the river bed, which consequently leave less room for the water to flow through, thus increasing its speed; and, of course, both waterfalls and rapids require a supply of water. When the river water level is high, rapids can actually end up lying underwater for a while.

Hraunfossar (Lava Waterfalls) is the name of a complex of waterfalls 1 kilometre wide in Iceland. The water flows over an impermeable layer, emerging from under an extensive lava field to flow via small waterfalls into the valley to join the River Hvítá.

Where does all the water come from?

The precipitation that falls as snow in the mountains melts as the temperature rises in spring, and this water forms small streamlets, which grow to become wild mountain streams. These streams meet many obstacles in their headlong rush to reach the sea, including deep ravines and hard bedrocks that force them to take a different course. Many waterfalls found in the mountains in the spring, are the result of the enormous increase in meltwater forcing its way to lower-lying regions and in doing so, pouring over steep rock faces. The summer months see most of these waterfalls drying up again due to lack of water. This is a seasonal phenomenon, producing few waterfalls that remain spectacular throughout the entire year. One exception to this is the Yosemite waterfall that falls some 739 metres in three stages and is fed by mountain streams

Heavy rainfall can really swell a waterfall, like these falls in Alberta, Canada. Here the water tumbles over a cascade.

from the Sierra Nevada in the United States. Another source of water arises when ground water springs from between rock layers. This kind of spring occurs when ground water cannot penetrate any lower because the underlying layers of rock consist of impermeable rock and the water-bearing layer (or aquifer) comes to the surface. Should there be a steep precipice at the spot where the water issues from the ground, the result is a waterfall or a series of waterfalls, such as the Hraunfossar Waterfalls in Iceland. Another example of a waterfall that arose in this way are the 'Roaring Springs' in the Grand Canyon. Rain and meltwater percolate into the ground on the northern side of the Grand Canyon and permeate through the bedrock until it reaches a porous layer of rock underlain by an impermeable layer, in this case Muav chalk overlying Bright Angel shale. As the water cannot permeate deeper into the ground it has to find another way, thus the water pours out of the bedrock along the shear plane of the permeable rock with the impermeable rock. At 'Roaring Springs', sufficient water pours out of the rocks all year round to provide the drinking water requirements of both the north and south sides of the Grand Canyon.

Some waterfalls are entirely dependent on the amount of precipitation that falls as rain. These are typically waterfalls in tropical areas, of which the Iguaçu waterfalls on the border between Brazil and Argentina are a good example. In the tropical rainy season the discharge is the greatest, whilst in the dry season the waterfalls may be completely dry, a situation that happens about once every 40 years. During a tropical rain shower, so much rain can fall that waterfalls arise spontaneously on mountainsides.

▷ *Offerufoss Falls, Iceland. The water thrusts its way downhill with torrential force.*

▽▷ *During a tropical rain storm so much rain can fall that rivers overflow and waterfalls appear spontaneously on mountain slopes.*

▽ *A difference in height is a vital requirement for the creation of a waterfall. This waterfall near Argentières, France, tumbles over a precipice.*

▽▷Firehole Rapids, Yellowstone Park, USA. The water from the Old Faithful Geyser flows into the Firehole River, raising the river temperature to high levels and giving the river its name. Trees that fall in are carried along by the river and can become additional obstacles.

2 River Basins

The life cycle of a river

As rivers evolve through time, the following stages can be distinguished. In the first stage, if a drainage system has not yet developed and nothing has cut down into the rocks, then lakes and morasses predominate. Small streams are not yet directly connected to each other. The following stage is when streamlets start to decrease, as a result of a lowering of the base level, causing them to start to join up with each other. A rudimentary drainage system evolves and the lakes and morasses disappear. Since the streams have a steep gradient at this point, they cut down deeply into the valleys and steep-sided, V-shaped valleys are formed. Erosion has scarcely affected the bedrock so that waterfalls and rapids are plentiful. Furthermore, a balance has not yet been established between slope wasting and erosion processes and the resistance offered by the bedrock. The longitudinal profiles of the rivers are just as varied as the sides of the river valley that, in part, consist of

The lowest stage of the Gullafoss Falls in Iceland. The water drops two stages here, that lie at right angles to each other. The water falls 32 metres into the approximately 70 metre deep, 2.5 kilometre long gorge of the River Hvítá.

Norway is rich in cascades. Here is a cascade of small waterfalls near Tromsa.

steep rock-faces. Ongoing erosion ensures that the rivers alter their beds. Variations in river beds are polished away; waterfalls and rapids disappear. The river's erosive powers almost cease; the valley slopes become less steep; the valley floor widens; the V-shape is lost and meandering bends form in the river. The final evolutionary stage is a wide flood plain with morasses and cut-off meanders, without much relief. The cut-offs have developed into oxbow lakes, fed by seasonal river floods. The longitudinal profile of the river now displays a gentle curve; this is called a balanced profile. In reality a river is a dynamic whole, which, together with the environment is trying to achieve some kind of equilibrium. Each situation is unique and many examples can be cited. Papua New Guinea and Exmoor, England, demonstrate two different examples of river formation.

Papua New Guinea: a profusion of waterfalls

New Guinea is the second largest island in the world; it lies to the north of Australia, just below the equator. The eastern part of the island is called Papua New Guinea. Various mountain chains define the relief of the island, with peaks often reaching higher than 4,000 metres. The mountains have steep slopes covered in thick forest in which a very rich flora and fauna flourish. The island is exposed for ten months of the year to rain-bearing winds from the sea, with the monsoon blowing from December to April from the north-west, whilst from May to September the trade winds blow from the south-east. When these moisture-saturated air streams meet the mountains of New Guinea, they rise, causing the warm, moist air to cool. This results in heavy rain that falls in a spectacle of thunderclaps and lightning flashes. Annually, more than two metres of rain falls, although in some places this may be as high as four metres. The rainwater gathers in small rivulets on the mountain slopes, and streams down the mountainsides through a network of gullies and steep ravines. Waterfalls can be observed from almost anywhere in the mountains and many of these waterfalls drop tens of metres. By the time the water reaches the lowland plains, the clear mountain streams are saturated with sediment, carried off the steep slopes in muddy torrents. As they meet the plains they turn into wide, muddy, chocolate brown rivers that snake their way across the flat plains to the sea, thus creating a pattern of river meanders. During heavy rains, floods occur on the lowland plains resulting in the transported silt being deposited on the plain. Some rivers flow too quickly to allow their sediments to sink, and so carry their sediment load out to sea. This can cause sediment clouds in the water for several tens of kilometres out to sea, which are visible from the air.

Exmoor, England: green country divided by fast flowing rivers

Exmoor lies in the south-west of England. It is renowned for its outstanding natural beauty, which includes wide expanses of heather and untouched fast-flowing rivers, which have a total length of 483 kilometres.

With the transport of sharp rock fragments the current scours the river bed, as is happening with this mountain stream in India.

Australia. In this stream the rocks have been partly scoured by stone fragments transported by the water.

▷The tumbling water works its way around a rock with a travertine deposit.

◁▽Rapids and waterfalls are great favourites with natural history photographers.

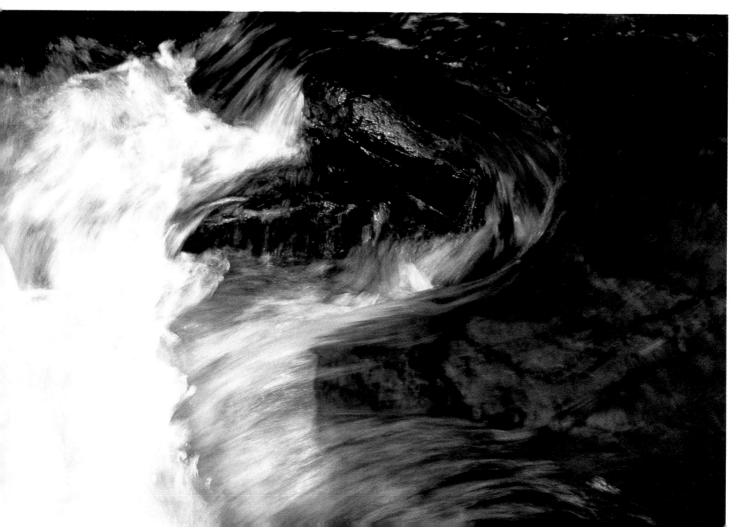

In this region of England, almost two metres of precipitation is recorded each year, making this a very wet area. The gently rolling, heather-covered moors close to the coast lie at a height of nearly 500 metres. The height of the heather-covered plateau causes the warm moist Atlantic air to rise, and cool becoming less able to absorb water vapour. The excess water falls as rain, snow and hail. The moors become so saturated with water in the wet season, which runs from September to May, that it gathers in shallow pools on the ground. Some of it percolates into the ground, 25% of it evaporates and the remainder trickles slowly over the surface to the edge of the moors, where it collects. This is how the River Barle is formed. In the first instance it is only an insignificant stream that flows over and under the heather to run into the Pinkworthy Pond at the edge of the moors. This is a small reservoir that was originally built in 1830. As more and more water collects in this pond it overflows, plunging noisily down over the boulders and pebbles of the river on its way. After a few kilometres the river loses its

Norway. Obstacles in the river, like the boulders that are just visible here above the water, cause rapids to form.

▷ *Mountain stream in the Haute Savoie, France. Some 25% of France's electricity requirement is met using hydro-electric power.*

restlessness and fast-flowing sections and its sandy banks interchange with deep, almost motionless pools. At Exe bridge, the crystal-clear River Barle joins the River Exe, which then flows south towards the sea.

The power of falling and flowing water

Contrary to common belief, falling water does have erosive powers. Raindrops, falling on bare ground, smash small particles of soil apart, leaving minute craters comparable to a meteor hit. One raindrop does not have much effect, but the

△▷ Rapids in the Bayerische Wald, Germany (above) and in Suonne, Finland (below).

number of raindrops is so large that together they cause a considerable amount of erosion. If the force of a falling mass of water is calculated, it appears to be capable of eroding rock. The force of the falling water from Niagara Falls on the underlying bedrock, which breaks the water's fall, lies in the order of millions of newtons. This is akin to dropping a block of rock weighing a thousand tons. Force like this creates depressions, called plunge pools, which are found at the base of waterfalls. The depression at the foot of Niagara Falls is 30 metres deep.

As the water falls it spatters into myriads of water droplets, creating a plume of mist that is often visible from a great distance. The plume of mist also ensures that there is enough moisture in the immediate vicinity of the waterfall for green mosses to grow.

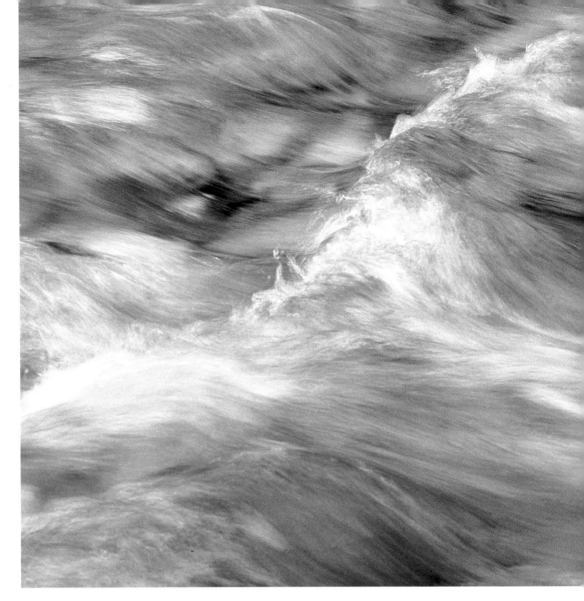

Standing waves are caused by obstacles on the river bed, like boulders. The water must either go over or round the obstacle. Immediately downstream of the obstacles, the water flows more rapidly and this increase in speed causes the water to create a hole. Close behind the hole, the water rises straight up in a seemingly motionless wave with the wave top curling upstream.

Firehole Rapids in the Firehole River, Yellowstone Park, United States.

Moss-clad stones in rapids. The stones are repeatedly moistened by the passing water.

The erosive action of a stream depends on how the water flows through the river bed and how much sediment is transported by the stream. If the speed is very slow, the water particles flow in parallel layers (laminar flow), but when the speed increases, the flow pattern is more complex and eddies occur, which are characteristic of turbulent flow. Generally the speed is high enough along the centre of the river to cause turbulent flow, and it is only along the sides and bottom of the river that the speed is slowed enough by friction to cause parallel flow. Increasing speed means an increase in the erosive action of the water as well.

The most common method of erosion in a clear river is the scouring of the river bed by the transportation of rough and angular pieces of rock. A spectacular example of this form of erosion is the creation of a pothole in hard rock. This happens in fast-flowing rivers where eddies form small hollows in the rock. If a large stone remains trapped in such a hollow, it is continually rotated by the turbulent flow. The rotating action of the stone wears a hole in the rock, a process called pothole drilling. The river uses existing faults and shear planes in the bedrock to deepen its existing bed, both by mechanical means, as described above, and by chemical solution of the rock along the fault or shear plane. The latter is particularly the case with limestone. The erosive action of the water is dramatically increased when the water also carries sediments, for the sediment particles have a scouring effect on the bed of the river.

If a landslide or a lava flow throws up a natural dam in the path of the river, the water simply gathers up behind the obstruction until it flows over the top and then begins to erode the dam. Waterfalls and rapids can be formed in this way. The speed at which a river flows through rapids is increased because the same amount of water has to force its way through a narrower gap. Continuing erosion of a waterfall slowly transforms it into rapids that are also finally broken down by the river.

Rock formed by water: travertine
Travertine is a rock that can often be found in combination with a waterfall or a spring. This rock is formed from calcium carbonate that has precipitated out and it occurs where lime-rich ground water meets the atmosphere. This is the case around springs and waterfalls in limestone bedrock areas.

Calcium carbonate is dissolved in the cold ground water. As the ground water springs from the ground, it comes into contact with the relatively warmer atmosphere where, because of the rise in temperature of the water, it is less able to carry carbon dioxide in solution. This causes carbon dioxide gas to be released into the atmosphere. This process is enhanced by the turbulence in waterfalls that causes the water to be in contact with the atmosphere over a greater surface area so that the gas can escape even more easily. As a result of the release of carbon dioxide gas, calcium carbonate precipitates out behind the waterfall. Mosses, which grow on the rocks in the damp habitat of a waterfall, take up carbon dioxide gas and intensify

Through reflection and refraction of the sunlight on the bottom, a blue-green colour appears. Together with ripples on the water surface, they cause dancing spots of light to appear.

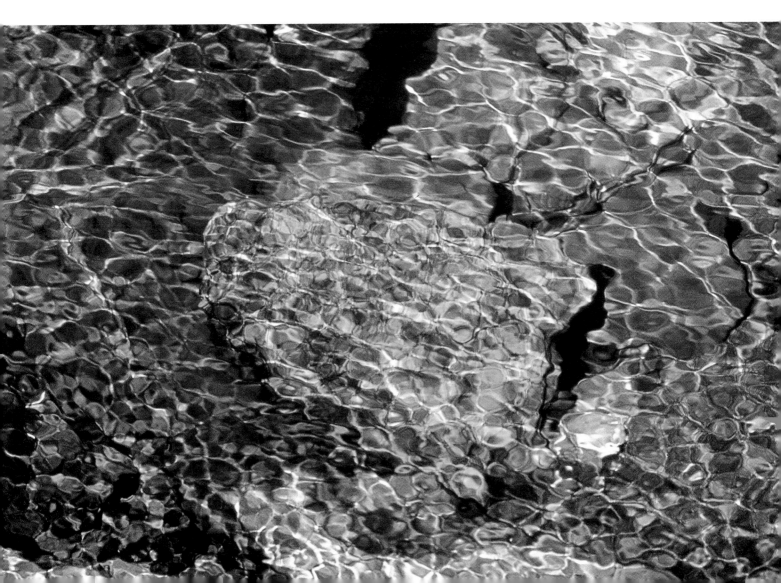

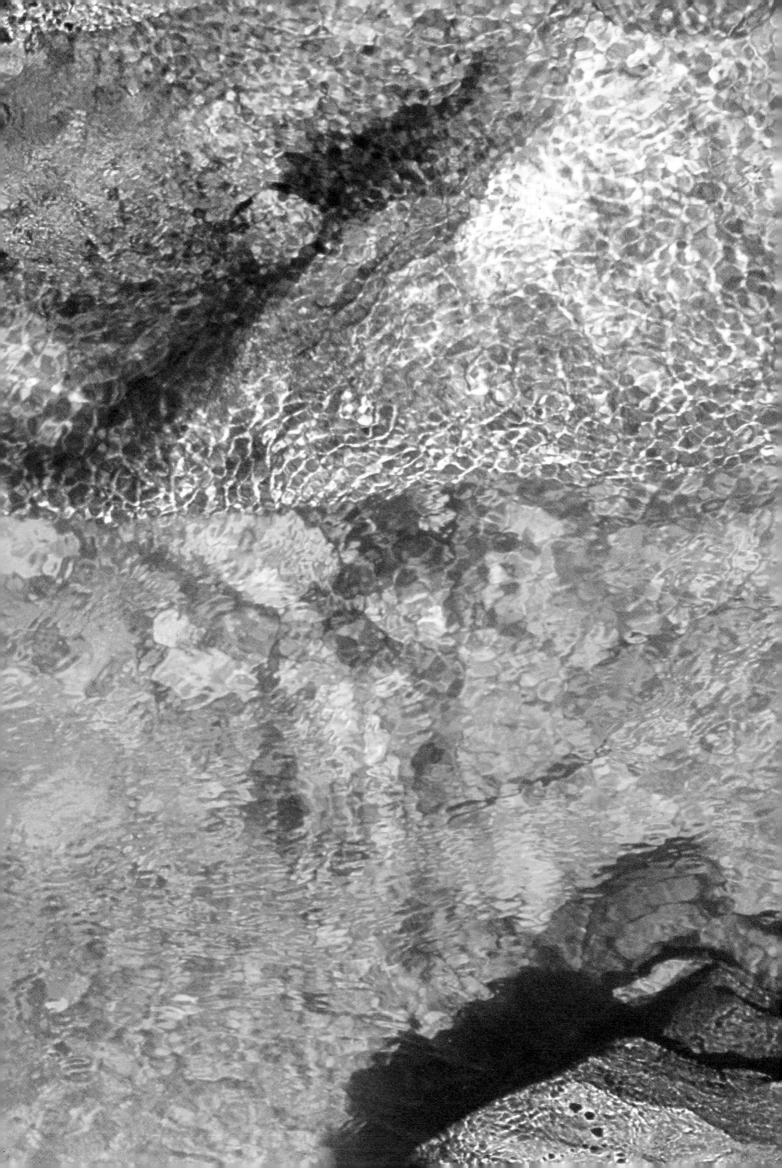

△Water pours off the Ceder Mountain in Algeria via ever more deeply, incised chasms. Chemical solution deepens the existing faults in the rocks. The waterfall will always choose the easiest course down.

▷These falls on Iceland have formed a natural bridge in the volcanic rock. The volcanic rock flowed into the river in earlier times, creating a natural dam. Instead of flowing over the top, the river undermined the lava flow thus forming a rock bridge. Weathering of the rock meant that the bridge collapsed in 1993.

△▷Softer rock underlies a more erosion-resistant layer. Both the percolation of water into the upper layer through cracks and the erosive action of the water as it tumbles down the cascade, serve to undermine the top layer and cause it to collapse. The waterfall continues to retreat, leaving behind a deep gorge. These are the Ram Falls near the Big Horn Dam in Canada.

Peyto Canyon, Canada. The water scours the river bed. The most common method of erosion in a river is the action of rolling and dragging along the river bed of sharp, rough fragments of rock. Through the rolling action the projecting rocks are gradually smoothed off.

Woodland stream valley on Exmoor, England

the deposition of calcium carbonate. If enough calcium carbonate has precipitated out, the mosses will petrify, and given that the structure of the mosses is retained and is porous enough to allow new mosses to grow on it, the circle of petrifaction repeats. The fine crystalline rock that arises in this manner is called travertine and, in the course of time, this forms a 'stone waterfall'. The calcium carbonate that has precipitated out, collects under the waterfall and forms a cone. The rock thus formed, can take on many shapes, of which the best known are stalactites and stalagmites. These are hanging (stalactites) and standing (stalagmites), mostly conical pillars that occur frequently in caves. In addition to fossilised moss structures, other plants may also be preserved. By studying preserved leaves and seeds, the paleo-vegetation of a region can be reconstructed and sometimes the age of the waterfall can be determined.

Geological structures that cause waterfalls

Most waterfalls are created as a result of differences in the erosion resistance of various rocks or by glacial processes that have left their traces in the landscape. Such differences in erosion resistance in a number of geological structures can lead, in conjunction with a drainage system, to abrupt changes in height, thus producing waterfalls. There are four clear examples. The first case is when hard, erosion-resistant rock overlies softer, less erosion-resistant rock. At the point where the river cuts into the soft rock it erodes away much faster than the hard, top layer, creating a difference in height. Another example of the formation of a waterfall can be found when there is a fault line. At the point where a hard bedrock lies against a soft bedrock, the soft rock erodes much faster. This is precisely

◁▽ *Pothole in solid rocks. A pothole arises in fast-flowing water in which eddies and whirlpools occur, forming small depressions in the stone. If a large stone comes to rest in the depression it is then continuously turned by the turbulent current. The stone drills a hole in the rock as a result of its continuous corkscrewing motion. This process is called pothole drilling.*

▽ *Standing water dissolves limestone in a river bed. Existing faults and planes in the rock are deepened by both mechanical erosion and chemical solution along the fault or plane. When the water is high the river can then erode the hollow even further.*

▷When the Grand Falls run dry, then sediment-rich water is left standing in small hollows in the rock. When the water evaporates only a thin layer of fine silt is left.

▽The waterless Grand Falls in Arizona, United States. On the right of the photograph, black volcanic rock can be seen, to the left, creamy limestone.

▷The speed and the sediment transported by the water, together leave scour marks in the limestone bed of the Grand Falls in the Little Colorado River. These are easy to see in the foreground. In the background lie the San Francisco Peaks, wreathed in thunder clouds.

what happened at the Victoria Falls and the Grand Falls in Arizona where the falls have developed along fault lines. Waterfalls may also be the result of the creation and withdrawal of glaciers that left behind U-shaped valleys with 'hanging' side valleys. Mountain streams tumble to the main valley floor from these side valleys. The structures formed by the glaciers were left behind in the last Ice Age, some 10,000 years ago. Finally, a large drop in sea level may also result in the creation of waterfalls. This is because the

The Ribbon Falls in the Grand Canyon (USA) are in the process of creating a beautiful travertine waterfall. Through the spattering of the lime-bearing water that springs from the ground just above the waterfall, carbon dioxide gas escapes causing travertine to be precipitated out. The plants petrify leaving behind calcified mosses. The travertine layer has grown so thick in the course of time that a pinnacle of travertine has grown under the waterfall.

Trebarwith, Cornwall, England.

base level of rivers is lowered, causing them to carve down into the bedrock in order to attain a state of equilibrium, once again creating waterfalls.

Collapsed limestone caves

The mountain range in Papua New Guinea, which already meets the most important criterion for many waterfalls because of its steep slopes, consists mostly of limestone. This rock type dissolves easily in rain water, which is slightly acidic. When the rain water percolates into the rock, tunnels and caves arise where the water gathers and flows further. This underground network can be very extensive in limestone regions, so that large rivers can be found flowing through underground tunnels. In addition to underground tunnels, huge caves develop, often tens of metres deep. Continuing percolation of water and solution of the rock cause the roof of the cave to finally collapse after becoming more and more porous. This creates a deep shaft into which a river can disappear several tens of metres into the ground in a thick curtain of water. The noise of the falling water echoes back and forth many times between the encircling walls of the shaft so that even the smallest stream can produce a thundering clamour. The water dissolves the underlying limestone forming a tunnel in the rock and flows from then on as an underground river. In Papua New Guinea there are many waterfalls that were created in this fashion. One example is the Nomad River on Mount Sisa, which, after a fall of several tens of metres, runs underground to reappear a couple of kilometres further away, on the surface.

Waterfalls that run dry

The Grand Falls in Arizona (USA) are the starting point of the canyon of the Little Colorado River, but for most of the year they stand dry. The Little Colorado rises in the Painted Desert and joins the Colorado River in the Grand Canyon after passing through a series of deep ravines. The river has developed parallel to various strike-slip faults. Along these sorts of fault lines, movement is almost exclusively horizontal, and this is why at Grand Falls, limestone has come to lie alongside volcanic rock. The volcanic rock is more erosion-resistant than the limestone, causing the river to retreat upstream forming a deep canyon in the limestone because of the continuous erosion. The river depends on a seasonal water supply from the Painted Desert, which gathers close to the Grand Falls before pouring into the canyon and forming the Little Colorado River. When the water supply is at its highest, thousands of litres of muddy water per second rush over the edge of the falls, leaving scour marks in the limestone bed. The fast-flowing water also forms potholes deep enough to hide a man, and forces its way into crevices where it undermines the limestone. This loosens the limestone and it tumbles down the falls to join the pile of huge boulders at the foot of the waterfall.

Waidbach, Germany. If fallen leaves become covered in travertine an imprint of the leaves is left.

Iceland. Undermining of the topmost lava layer frequently causes boulders to crash down the falls.

In order to reach the other side of the river a swaying rope bridge has been slung over the waterfall. The yellow colour of the water is due to the sediment carried by the river. These falls are part of the Iguaçu Falls in Brazil.

The sediment-filled brown waters of the Polousse Falls, United States, plunge over a hard top layer. A recess in the rock wall has been created by the continuous erosion. The erosion is increased by the scouring action on the river bed of the sediment transported by the river.

3 World Famous Waterfalls

Niagara Falls

There are about 100 waterfalls in the world that are higher than Niagara Falls and at least one that has a greater volume of water, but few are more famous. Until about 10,000 years ago, enormous glaciers covered a large part of North America and forced the water of the Great Lakes south, towards the Mississippi River. As the glaciers melted at the end of the Ice Age, the Niagara River was formed. It drained from Lake Superior, Lake Huron and Lake Michigan via Lake Erie, before it crossed a small plateau and dropped over a steep precipice into the lower-lying Lake Ontario. Today, the Niagara River forces its way past anything that obstructs its 55 kilometre path with thunderous power, pounding the horizontally stratified rocks as the water plunges over the Falls. This bedrock is part of a geological structure known as the Niagara Slope. The upper rock layers of this ridge consist of hard Lockport dolomite. It is often thought that the raging waters at the foot of the Falls are undermining the overlying rock layers, but that is only partly true, for the process of undermining is much more subtle. At the foot of the Falls, a 30 metre deep hole has been carved out by the action of the tumbling water; but the overlying rock layers are not being undermined under the Falls, but rather above the waterfalls. Here, the water seeps into cracks in the dolomite, percolating down and dissolving the underlying layers of slate and sandstone. This causes the dolomite to collapse and break off in plates that are sometimes 50 metres in length. The waterfall probably reached its present position when Goat Island divided it into two different cascades. To the east of this wooded island, on the United States side, lie the American Falls, 56 metres high and 323 metres wide; these carry less than 10% of the water of the Niagara River. The rest of the water pours into Canada over the Horseshoe Falls, 54 metres high and 675 metres wide, which, as the name suggests, are shaped like a horseshoe. More than five million litres of water per second tumble over the edge. The bedrock is eroding more rapidly on the Canadian side because the greater volume of water pours over on that side, so that the Horseshoe Falls have retreated some 300 metres in 300 years.

Today, some of the water is diverted to generate hydroelectric power and the Falls are retreating at about 30 to 60 centimetres per year. The speed of the river current at the Falls lies between 25 and 30 metres per second. Erosion is progressively moving the waterfall upstream, leaving behind a deep ravine, some 90 metres wide. At about 4.8 kilometres downstream from the falls, the river turns sharply to the right, forming a whirlpool that leaves the river unnavigable. All the water that flows over Niagara Falls, eventually finds its way to the Atlantic Ocean after its long journey to the coast. Many stories owe their origin to the Niagara Falls. One famous tale concerns Captain Matthew Webb, the first man to swim the English Channel. On 24 July 1883 he attempted to swim across the ravine below the falls, but he underestimated the strength of the current and drowned in the whirlpool.

The fame of Niagara Falls grew through spectacular feats of daring, and in 1859 they even attracted the famous French acrobat, Charles Blondin (1824–1897) to the site. Blondin hoped that all previous stunts would pale into oblivion compared to his crossing of the falls on a wire in four minutes. After his successful crossing, he then amazed the public by repeating the crossing over the roaring falls several times. He crossed once whilst blindfolded, and then followed this by crossing with his impresario on his back. Next, he carried a small stove to halfway across and fried an egg, and as a finale, he walked across on stilts. Niagara Falls have not only drawn writers and daredevils: their spectacular beauty has also drawn romantics and honeymoon couples. The first couple to marry at the Falls were Jerome Bonaparte, the younger brother of the French emperor, and his bride Elisabeth Patterson in 1803. They were followed by many couples,

Niagara Falls actually consists of two waterfalls, of which the Horseshoe Falls on the Canadian side are the most famous (on the right in the photograph). The horseshoe shape, to which they owe their name, is clearly visible. On the left of the picture lies Goat Island that divides the Niagara River into the two falls.

establishing a custom that drew the following cynical quip from the Irish playwright, Oscar Wilde (1854–1900): 'Niagara Falls is the second, big disappointment for an American bride.'

Victoria Falls.
The Zambezi River flows over a wide basalt plateau before plunging into a narrow gorge, nicknamed the Devil's Cataract, and continuing on its way zigzagging through a series of gorges. On the west side of the Falls (on the left of the photograph) the river is busy changing course backwards by cutting a new gorge along a fault line in the Karoo basalt.

Victoria Falls

Victoria Falls lie on the Zambezi River on the border between Zimbabwe and Zambia, 140 kilometres upstream from Lake Kariba, in Africa. From a vast plain, the waters of the Zambezi River pour over a 1,700 metre wide steep rock face to plunge 108 metres into a narrow gorge. These falls are more than twice as wide and twice as high as Niagara Falls. On average, 935,000 litres per second pour over the edge into a gorge that is 75 metres wide. The local Kalolo tribe call the waterfall Mosi-oa-tunya, the

When sunlight reflects through the falling water a rainbow appears. If a beam of sunlight is reflected in a drop of water, it is split into a continuous colour spectrum, that appears as a rainbow. The plume of mist from the Victoria Falls can be seen from 40 kilometres away.

'thundering smoke', on account of the plume of mist that rises 300 metres into the air and is visible from a distance of 40 kilometres. The river is dependent on rain for its supply, so the waterfall can run dry. Zambia uses the energy of the falling water to generate hydroelectric power. Below the falls, the river runs through a gorge that zigzags for several kilometres, a remnant of the continuous erosion of the Victoria Falls. The zigzag pattern arose from fault lines that cut across the river bed, which were easily eroded. The Zambezi River flows through Mozambique to finish up discharging into the Indian Ocean.

The Iguaçu Falls

The shattered edge of the immense Paraña tableland in southern Brazil produces the natural conditions necessary to create a number of waterfalls. The Iguaçu Falls, which lie on the border with Argentina in the province of Missiones, are generally recognised as the most impressive. Iguaçu means 'powerful water' in the language of the local Guirani Indians. The mighty waterfall of Iguaçu consists of 275 individual cascades, each separated from the others by tree-covered rocky islets. Some falls tumble unhindered from the 4 kilometre wide edge straight into the 82 metre deep ravine, whilst others fall in a series of short cascades from ledge to ledge. The structure of the bedrock of the tableland splits the waterfall up into innumerable streams. The bedrocks of the tableland consist of solidified lava and basalt,

The Iguaçu Falls in Brazil cascade over the edge of a basalt plateau, just like Victoria Falls. They plunge over a width of 4 kilometres via numerous steps into the Devil's Throat some 64 metres below.

In the rainy season the Iguaçu Falls reach their full glory, when more than seven times the volume of water of the Niagara Falls thunders over the edge.

erosion-resistant volcanic rocks. Natural differences in height occur due to the irregular structure that has resulted from many volcanic eruptions. Islands of basalt force the river to split up into several channels. The rise and fall in the water level in the Iguaçu River is almost solely dependent on the seasonal rainfall in the catchment area. At the height of the rainy season, from November to March, the river undergoes a massive expansion and an estimated volume of 12,768,750 (nearly 13 million) litres of water per second plunges over the Iguaçu Falls; this quantity is more than the contents of six Olympic swimming pools. Eleanor Roosevelt (1884–1962), wife of the

▷The jungle round the Iguaçu Falls harbours many rare species of animals, like the solitary jaguar and the Brazilian tapir. The latter is a good swimmer and diver, feeding at night on a diet of leaves, fruits and water plants. In this region there are some two thousand species of plants, about four hundred different birds and innumerable insect species.

American President Franklin D. Roosevelt, once said, 'After seeing the Iguaçu Falls our own Niagara Falls looks like a kitchen tap.' The Iguaçu Falls, at their greatest, are more than four times wider, one and a half times deeper and carry more than seven times as much water as Niagara. Within the catchment area of the Iguaçu River, rainwater flows directly into the river over the ground that man has cleared of trees. The amount of water is therefore very irregular and it alternates between flooding and shrinking to a fraction of its greatest size. In 1978, the cascades dried up completely and the thunder of falling water was muted, until four weeks later, when the first streamlets began to appear. Similar periods of drought only used to occur about once in 40 years. Nowadays, the construction of a dam upstream means that man can regulate the supply and the amount of water passing

over the Falls according to the energy demand. The waters of the Iguaçu River plunge into a 80 metre deep, narrow gorge called the Garganta del Diablo, 'The Devil's Throat', and then wind their way to meet the Paraña River some 22 kilometres downstream. The Paraña is the third largest river in South America, after the Amazon and the Orinoco, and rises in the north of the country, flowing into the Southern Atlantic Ocean near Buenos Aires. In 1541, the Spanish explorer, Alvar Nuñez Cabeza de Vaca had the honour of being the first European to see the Iguaçu Falls, whilst searching for the treasures of the Incas. He gave the Falls a Christian name but this was rapidly replaced by its current name, which was used by the local Guirani Indians. The extremities of the U-shaped Iguaçu Falls are marked by the Union Falls and the San Martin Falls that lie opposite each other. Between these two gigantic falls, many smaller waterfalls, including 'The Three Musketeers' and the 'Belgrano', plunge into the gorge over rocky ledges and overgrown terraces.

In his eulogy to the Iguaçu Falls the Swiss botanist, Robert Chodat, also described the rich flowering flora, '... an exuberant, almost tropical vegetation, the leaves of large ferns, the bamboo stems, the graceful trunks of palm trees and a thousand different sorts of tree, their crowns bending over a bay decorated with mosses, pink begonias, golden orchids, brightly coloured bromeliads and lianas with trumpet flowers...'. The National Parks that Brazil and Argentina established on their respective sides of the waterfalls, protect the rich tropical and sub-tropical wildlife. In the parks, some 2,000 species of plants and more than 400 species of bird can be found. Birds such as tinamous and parrots inhabit the trees, whilst swifts nest in the rocky crevices of the waterfalls and swoop low over the river, feeding on the huge swarms of insects. Ocelots and jaguars roam through the undergrowth, as well as tapirs, deer and peccaries. On the rocky islands that split the water into many channels, many different varieties of trees grow, such as the cedar (Cerdella fissilis), trumpet liana and lapacho. A number of unusual water plants of the Podostemaceae family flourish on the ledges of the waterfalls; these are flowering plants although they resemble mosses and lichens. Such is the specialist niche that this family of plants occupies that they are only found near turbulent or cascading water.

The Yellowstone River
It is difficult to find words to describe the breath-taking splendour of the world's first national park. This unique, beautiful and ever-changing landscape in the United States overwhelms the visitor with its wonders. Nowhere else on earth does the fluid heart of the earth beat so close to the surface as here, with mud, hot water, steam and fumes, bubbling and spluttering from every pore in the ground. In addition to world-famous geysers, like Old Faithful which spouts hot water every hour, this park also harbours a number of splendid waterfalls in the Grand Canyon of the Yellowstone River. The Yellowstone River rises in the Absaroka Mountains in the southeast of the park, and flows northwards, carrying rain and meltwater from the eastern side of the continental divide. The Grand Canyon of the Yellowstone River is a narrow, deep gorge, which in the last 500,000 years has seen various periods in which it was alternately in-filled

or eroded. Ash and debris from various volcanic eruptions have repeatedly filled the valley. After each episode the Yellowstone River then once again carved out its valley in the volcanic deposits. Glaciers and icecaps also blocked the river several times, with the last icecap melting away about 12,000 years ago. Since then, the Yellowstone River has eroded away most of the volcanic and glacial deposits and carried them out of the gorge. The sides of the gorge are cut down through rhyolitic lavas and tuffs. Rhyolitic lavas are volcanic lava flow rocks with a very high silica content (c. 70%). They are very viscous when hot and, after cooling, they are just as hard as quartzite. Intensive chemical erosion by hot ground water has changed the colour of the rocks from an unweathered grey-black into a sparkling red, pink and yellow. It is to these 'yellow stones' that the park owes its name. Close to the foot of the gorge where hot springs bubble up and steam rises from the ground, hydrothermal processes have altered the composition and characteristics of the rocks. The Upper and Lower Falls lie at the start of the deepest and most spectacular section of the gorge. The Yellowstone River drops 94 metres at the Lower Falls, over the edge of resistant rhyolitic lava that erodes more slowly than the underlying volcanic tuff and the chemically weathered lavas. The Upper Falls are approximately 33 metres high and lie upstream from the Lower Falls. At the end of the Grand Canyon of the Yellowstone River, stand Tower Falls, where Tower Creek plummets over the edge of a hanging valley to land in the Yellowstone River some 40 metres below. Tower Falls are thus called because of the tower-like rocks that stand sentinel over the Falls. The rocks were shaped through water forcing its way into cracks in the rocks and widening the cracks by dissolving the rocks.

The Firehole River is warm, in comparison to other rivers in Yellowstone Park, because of the many geysers and warm springs that drain into it. This is why, in the early morning, a white blanket of mist hangs over the river.

Yosemite: snow and water flow over granite rocks

Yosemite National Park is a wonderful area in the mountains of the Sierra Nevada in California, in the United States. It is a landscape full of high mountain tops, thundering waterfalls, gigantic trees, effervescent rivers and calm meadows. The highest waterfalls in the United States lie in the park, which was established in 1864. The name, Yosemite, is the English corruption of Uzumati, meaning grizzly bear, which was the name of the Indian tribe who originally lived in the area. The Sierra Nevada mountains consist of a complex of massive intrusive rocks. The geological processes that have shaped the landscape, are erosion, stream erosion, slope wasting and the formation of glaciers, but the uniqueness of the topography was determined by the existing structures and fault systems in the bedrocks. When the mountains were thrust upwards, the erosion of the overlying rocks and the resulting reduction in pressure caused fissures to open in the bedrocks. These fissures, along with the incision of rivers associated with elevation and Pleistocene glaciers, have all had an influence on the development and location of the waterfalls in the Yosemite. These are:

- Yosemite Falls 739 metres
- Ribbon Falls 491 metres
- Staircase Falls 396 metres
- Bridal Veil Falls 189 metres
- Nevada Falls 181 metres
- Illilouette Falls 113 metres
- Vernal Falls 97 metres

The Yosemite Falls, with a height of 739 metres, fall in three cascades, and are the highest waterfalls in the United States. The discharge from the Falls is greatest in the spring because they are mainly fed by meltwater from higher up the mountains. The regional elevation and tilting of the Sierra Nevada mountain range to its present height, disturbed the former drainage system of the whole region. In the later phases, the elevation was so fast that the resulting steeper gradient in the main rivers caused the valleys to erode more rapidly. This erosion was much faster than in the side valleys where the elevation had almost no effect on the gradient. By the time of the start of the last ice age, some 30,000 years ago, the side rivers were already plunging via steep cascades into the main valley, from several tens of metres above the valley floor. When the glaciers filled the valleys, the erosive action of the ice deepened the main valley and scoured its sides, leaving them even steeper and accentuating the 'hanging relationship' of the side valleys. Many cascades were thus transformed into free-fall waterfalls. The Yosemite, Ribbon, Bridal Veil and Illilouette Falls all arose in this fashion. The Nevada and Vernal Falls have a different geomorphologic history; they arose during the ice age. The Merced River flows over a gigantic glacial staircase, which was formed when the ice moved into the valley, producing the Nevada and Vernal Falls. During this ice movement, huge pieces of broken granite were carried along, but these massive blocks were left behind when the ice retreated, forming a glacial staircase with steps of 100 metres or more in width.

◁*Amidst high-rise rocky pinnacles, Tower Creek cascades via Tower Falls into the valley of the Yellowstone River. A winding path brings tourists to the foot of the falls. The rock walls around the falls are a breeding site for the White-throated Swift.*

The Yellowstone River drops via the Lower Falls into the V-shaped Grand Canyon of Yellowstone. The normally sparsely vegetated gorge is turned green, within the range of the spray from the waterfalls, by mosses that grow there. Further away from the water vapour, lack of water prevents them from growing.

The sides of the gorge that the Yellowstone River flows through are coloured red, pink and yellow by the chemical action of hot steam escaping from the ground. The sides are always exposed to slope wasting so that landslides are a frequent occurrence along the river banks. This makes it difficult for trees to grow here.

The Yosemite Falls are the highest in the United States. They drop 739 metres in three cascades. The uppermost cascade is 435 metres high, the middle cascade 205 metres and the bottom cascade, a mere 97 metres. The falls flow from a side valley higher up into the main valley of the Merced River. This high, hanging valley was created when the valley of the present Merced River was eroded by a glacier.

Natural barriers

Rapids and waterfalls in rivers are natural barriers that hinder commercial navigation. If it is necessary, for the economy of a country, to navigate a river, despite the presence of rapids, then man generally has a way of solving the problem. For a long time the Trollhättan Falls in the south of Sweden formed a barrier to shipping between the enormous Lake Vänern and Gothenburg. In 1718 a series of eight locks were built around the waterfalls, but it was only in 1832 that the project was completed and a canal had been constructed with 65 locks. Now it is possible to sail between Gothenburg, via Lake Vänern and Lake Vältern, to the Baltic Sea port of Mem. The total distance between the two cities is 560 kilometres of which 87 kilometres consists of artificial waterways. The highest point lies 91.5 metres above sea level.

The spray clouds of the Lower Falls, in Yellowstone National Park, hide the foot of the waterfall.

China: a secret world behind the waterfall

Hidden away in the south-east of China, in the west of the province of Guizhou, lies the town of Huangguoshu, a veritable waterfall paradise. Just to the north-west of this town lies the source of the river Beipan. This river produces no less than 18 waterfalls around Huangguoshu, including the Huangguoshu Falls, which at 74 metres high and 81 metres wide, are the highest in China. What is unusual about these falls is that behind them lies a 134 metre deep cave that can be visited. In the 'Curtain Cave', as it is called, you are virtually overwhelmed by the thundering roar of the water. Most waterfalls here fall over lava plateaus, which can be found in a great variety of shapes and forms.

The Angel Falls

Angel Falls, the highest falls in the world, lie in the south-east of Venezuela in a region characterised by table mountains that rise two kilometres above the surrounding countryside. The rocks of the table mountains are 1.5 billion years old and have been sculpted by weathering processes into a maze of ravines. In these ravines during the wet season (from April to November), myriads of waterfalls plummet to the floor of the valleys; one of these is the Angel Falls, which could be considered the goddess of waterfalls. The local Pémon Indians call it Kerepa-kupai-merú; meaning 'falling from the deepest place'. It plunges unbroken from the mountain Auyán Tepui for a distance of 807 metres. Further cascades bring the total length of the falls to 979 metres, some 20 times higher than Niagara Falls.

The Huangguoshu Falls (left) are the largest in China. Behind the falls lies a 134 metre deep cave. Other waterfalls can also be found in the vicinity of Huangguoshu including the Luositan Falls (right). They spring from under a lava field which overlies an impermeable layer. They are the highest falls in the Huangguoshu region.

The Pémon Indians claimed that the Falls were the result of a battle that raged between the two gods, Canaima (the devil) and Cahuña (the good god). After a hard battle, Canaima won and the region of the waterfalls became his. Cahuña cried so bitterly that his tears caused the waterfall to fall out of the sky. The legend is reflected in the local names: Auyán Tepui means 'Devil's Mountain' in the language of the Pémon Indians. The mountain is, in fact, almost impenetrable with deep ravines fringed with dense jungle. If one wishes to see the waterfalls today, there is a choice between a flight in a small aircraft over the falls or an arduous, two day trek over the fast-flowing Chúrun River. The Chúrun River flows through a gorge called the Devil's Gorge, which cuts the plateau of Auyán Tepui in two. The roar of the Falls can be heard from a couple of kilometres away.

◁△▷ This spot is called the Devil's Bathtub. In the spring, meltwater streams flow over this steep rock, down into the glaciated valley. In the course of the summer, the meltwater mostly dries up and only thin trickles of water run down the rock face. Mountaineers often climb up from the floor of the valley to rest at this point against the cooling, refreshing wall of water.

The Iguaçu Falls viewed from the triple border point where Argentina, Brazil and Paraguay meet.

The Rhine starts life as a small mountain stream in the Swiss Alps. Close to the border with Germany, the Rhine is already large enough to create Europe's best known waterfalls, the Rhine Falls at Schaffhausen.

Angel Falls were unknown to the outside world before 1935. Spanish gold seekers had searched for the Falls because, according to the Indians, gold could be found where the god Cahuña cries, but they never found it. In 1910, Ernesto Sánchez took an expedition into the region and this resulted in the Falls being marked on maps, even though he never found them. The real honour, however, fell to the pilot, Jimmy Angel, after whom the Falls are named. He discovered them in 1935 whilst flying on an expedition to explore the Guyana Highlands in eastern Venezuela. A foot expedition in 1949 confirmed that the falls were indeed the highest in the world.

Iceland: land of waterfalls

Iceland is not only a land of volcanoes, geysers and glaciers, but also of many delightful waterfalls. Its geographical position and geology are ideal for the emergence of waterfalls, since, as a result of the many volcanic eruptions that Iceland has witnessed, there are many lava plateaus. These plateaus generally end abruptly, so that a steep drop to older, lower-lying lava plateaus is the result. In some places, several steep precipices occur one after the other, like a series of steps, thus forming a ladder of waterfalls. Given that the layers of lava are never very thick, no more than a few tens of metres, really high waterfalls are a rare commodity in Iceland. One of the better known falls, the Godafoss, are only 12 metres high and at least 100 metres wide. This lack of height is more than compensated for by the plethora of waterfalls that can be found in a small area. Large quantities of precipitation fall on Iceland so the rivers always carry lots of water, and because of the high river levels the waterfalls are often very wide. Iceland also has a number of wonderful phenomena that are associated with waterfalls. There is, for instance, a waterfall that eroded through a layer of volcanic rock and created a natural bridge above the waterfall. This left the rock that formed the bridge exposed on all sides to the effects of weathering. As a result of the action of frost, whereby cracks and fissures in the rock were widened by the process of freezing and thawing of the water lying within them, the cohesiveness of the rock slowly dissipated. This resulted in the collapse, in 1993, of the natural bridge above the waterfall.

In mountainous regions like the Alps, the temperatures can fall so low in the winter that even waterfalls freeze up. The most daring and dangerous branch of mountaineering is the ascent of frozen waterfalls during which the climber cannot be secured. One wrong step could easily be fatal.

▽▽The Godafoss Falls in Iceland are only 12 metres high. These wide falls owe their name to the fact that the head farmer and magistrate Thorgeir threw images of pagan gods from his temple into the Falls following a decision in 1,000 AD to adopt the Christian religion.

▷The Skogafoss Falls are one of the highest falls in Iceland. The Falls are 60 metres high, and according to an old legend a chest of gold lies hidden at the foot of the falls.

The water in these falls in Iceland looks like a veil draped over the boulders. This effect was obtained by using a long shutter time when photographing the falls.

Another very unusual waterfall is the Hraunfossar, which is not actually one single fall but a whole complex of waterfalls, whereby the water flows out of the valley sides of the Hvíta River. The water reaches the surface in the valley sides because the impermeable underlying rock layers, over which it flows underground, come to the surface here. The result is a series of waterfalls that, over a length of one kilometre, flow into the Hvíta River.

Icelandic history is rich in sagas about waterfalls and some have given the waterfalls their name, like the Godafoss Falls. The story goes that the head farmer and magistrate, Thorgeir Thorkelson of Ljósavatn (a farm to the west of Godafoss Falls), on returning home from the annual public legislative meeting in the year 1,000, took the images of the pagan gods out of his temple and threw them in the waterfalls. The meeting had decided to adopt the Christian religion. According to an old legend, a chest of gold is hidden under the Skogafoss Falls. Whether one has to look for it at the end of the rainbow, that shimmers in the spray of the Skogafoss Falls in good weather, is, regretfully, not revealed in the story.

4 Flora and Fauna

Natural selection

The mechanical force of a fast-flowing river has a tendency to rip away the plants in the current and to drag animals along with it, resulting in a drastic selection taking place with only those species surviving that have best adapted themselves to the circumstances. The greatest diversity of species is generally found downstream in a river where the water is warmer, and along the riverbanks where the current is slower. In such places the survival chances for the various inhabitants of the river are the highest. Apart from the speed of the current and the temperature of the water, there are a number of other factors that can limit the occurrence of plants and animals. These factors include differences in the nature of the soil, turbidity, oxygen content and the amount of organic material in the water. Since these factors vary from the source to its final discharge into the sea, the whole course of a river can be divided into different zones of flora and fauna.

Dussart (1966) divided the river up into various biological zones according to temperature. These zones can move if the amount of water being discharged by the river alters. This change in quantity can increase the turbidity of the river or raise the temperature, with sometimes fatal consequences for certain species. There are organisms that can shift temporarily from one zone to another. Certain species of fish swim upstream during the spawning season to breed, whilst other fish species that are lake-dwellers, head in precisely the opposite direction to the sea to spawn. In both instances, zonal limits are crossed. Out of the spawning season the various species of fish prefer to keep to certain parts of the river, where temperature plays an important role.

Fauna in cold flowing water

The species of fauna that are found in the coldest zones of the river have a much greater affinity in some way with rapids and waterfalls. These zones usually lie in mountainous regions, which is where, for instance, trout and salmon can be found, and where river shrimps and the larvae of dragonflies and caddis flies abound. All these species are able to cope with temperatures below 15°C. At the source of a river, the number of species and the numbers per species are both at their lowest. The organisms that survive here, such as the larvae of the caddis fly, are only found between the rocks or on sandbanks because the current is not so strong there.

Some distance downstream lie the spawning grounds of salmon and trout, both of them species that are used to living in cold, turbulent water. The turbulence ensures that the water is very rich in oxygen. A little further downstream the turbulence is less, because the river bed is less steeply sloped. This zone can usually be found at the foot of the mountains. Rapids and waterfalls can still be found here, but in between these features the

▷ On rock faces around waterfalls, the type of vegetation mostly commonly found is mosses. Waterfalls are an important source of nutrients for them.

water is much more placid. The temperature in this zone of the river is also slightly higher.

In this zone, species like the gudgeon are found, a species of fish that likes quiet, clear water with a muddy bottom. These quieter parts of the river are also home to the river shrimp. This creature lives on organic material that

The tundra climate that prevails in Iceland means that the vegetation on the island consists mostly of mosses, which seem to cover the landscape like a green blanket.

drifts over the bottom of the river and uses calcium salts that are in solution in the water to strengthen its protective plates.

All this animal life in the upper reaches of the river also draws a number of bird species, which have specialised in looking for food in fast-flowing water. The Harlequin Duck from North America and the shelduck from South America spend the entire summer high in the mountains on fast-flowing rivers, feeding on insect larvae and raising their young before the start of winter. In the winter they leave the mountains and settle at sea or on lakes. The appearance of ducks

Grizzly bears abandon their territorial instincts during the annual salmon run upstream. The photograph above shows a group of grizzly bears waiting for the fish at a waterfall. Large groups of bears frequently gather at points like these in the salmon season.

When the salmon run is at its height, the effort of catching a fish is quickly rewarded. The salmon that this bear has caught is some 50 centimetres long and contains the animal protein so badly needed by the bear after a long, cold winter.

in fast-flowing rivers is not quite as remarkable as the story of the dipper. This bird, which is related to the wren, has adapted extremely well to life in and around flowing water. This little bird, barely 20 centimetres in size, moves using strong wing beats through the flowing water, picking up insect larvae between the rocks or from the river bed. Its nest is often under or near a waterfall and whilst it may seem strange, this choice of nesting site is not so unusual. Species like the Yellow Wagtail and the Black Swift also prefer to nest close to a waterfall. Most birds would normally be frightened off by the thunderous roar of the cascading water, which prevents them noticing approaching danger. High-flying birds like kites are another exception, making use of the thermals that arise above a waterfall. When a waterfall plunges into a narrow ravine, a rising column of air occurs on which the birds can easily gain height.

Plants that defy falling water

Although these examples are a good indication that many organisms are at home in and around waterfalls and rapids, the adaptation is not as extreme as that of the plant family Podostemaceae. This is a family of water plants that occurs in South America, especially around the Iguaça Falls and in fast-flowing tributaries of the Amazon. The family includes species that grow in the mists of the waterfalls and under rocky ledges. There are even types that defy the tumbling waters and grow directly under the waterfall. The Podostemaceae are highly sensitive to changes in temperature and all members of the family only grow in or close to fast-flowing or falling water. In these positions they are continually moistened by the falling water droplets, leaving them unaffected by the tropical heat that predominates in the jungle. The plants, that are a bit like mosses and lichens, cling to the rocks by means of suckers. They flower in the dry season when the river is low. Seeds are then formed that land on the bare rocks. The seeds cling to the rocks with tiny suckers and germinate only once the water rises again. This way of life has remained unchanged throughout evolution.

The waterfall as a source of food

Rapids and waterfalls form a natural obstacle in the course of a river and are the reason that food gathers at that point. A good example of this is the relationship between waterfalls, salmon and grizzly bears in Alaska. Every year, salmon swim many hundreds of kilometres to return to their spawning grounds to breed. During their trip they are repeatedly confronted with rapids and waterfalls. The salmon pass these obstacles by jumping up out of the water. They are then very vulnerable to predators. Grizzly bears, like man, like salmon a lot, besides which the salmon provides the grizzlies with sorely needed, high protein food. However, salmon are not easy to catch when they are swimming freely, but traversing rapids brings the salmon into shallower water where they form an easier

Some water plants that closely resemble mosses and lichens only grow in fast-flowing water. The flowers of these plants, including Dicraeia algiformis, bloom at the end of the rainy season. They belong to the Podostemaceae family. The seeds that are scattered during the dry season only germinate when water flows over them again.

catch for the bears. The bears have learned to toss the salmon onto the river bank when they jump out of the water. Given that salmon take some time to pass through rapids, a lot of salmon gather there and the chance of success for the bears is higher. A good fishing spot will draw tens of bears at any one time because, as food is so plentiful, there is no real competition.

Other animals are also drawn by the salmon run. Seagulls fly inland during the spawning period of the salmon to feast on the remains left by the bears or to catch a salmon themselves. After the spawning period the salmon die in their thousands and their carcasses lie in the river rotting, leaving the nutrients to be absorbed by the fast-flowing water.

Waterfalls, rapids and man

The influence of waterfalls and rapids on man has two aspects. On the one hand, energy can be generated from fast-flowing and falling water and on the other hand, too strong a current impedes the navigability of a river and makes building bridges, for example, extra difficult and hazardous.

Shooting the rapids

Shooting the rapids is one of the most physically challenging water sports and it is practised almost everywhere in the world. In mountainous regions, it is a magnificent challenge in the spring to navigate the mountain streams, which are swollen with meltwater. This is done in kayaks - small, narrow and highly manoeuvrable craft - which are built to be able to take the rapids without a problem. To prevent people getting themselves into life-threatening situations, a world-wide classification has been established that categorises rivers and streams into six classes, from very easy to unnavigable. In Europe all categories are represented. The Alps and the Pyrenees are home to many rivers where the sport of kayaking can be practised. In addition to kayaking, rafting on white water rivers is a sensational experience shared by several people in one boat. Usually between six and eight people share a rubber boat.

The Colorado River is busy in the summer months with rafts; about 200 people a day make the trip from Lake Powell, through the Grand Canyon to Lake Mead. When shooting the rapids, whether one is in a small kayak or in a larger rubber boat, the art is to slide down on the tip of the tongue of water. This is a piece of smooth water that is V-shaped, of which the narrow part points downstream; in this section there are no waves, eddies, rocks or other obstacles. The tongue can extend over the entire width of the river. Whenever there is no tongue, there are a number of possibilities: the boat can be 'ported' down past the rapids if there are enough people, or it can be lowered through the rapids with ropes attached to the bow and stern. The last possibility is to abandon the boat and continue the trek on land. In rapids the boats achieve high speeds: speeds of 35 kilometres per hour are not exceptional. At these sorts of speeds the ability to react rapidly is vitally important to prevent the boat crashing onto the rocks or being drawn into a whirlpool.

Great waterfalls are real tourist attractions, like the Lower Falls in Yellowstone Park, USA, which are shown here. These famous Falls draw millions of visitors each year.

◁△ Rapids are a real challenge to white-water sportsmen and women. To be able to face the challenge of the white-water rivers, good technique is just as necessary as good equipment.

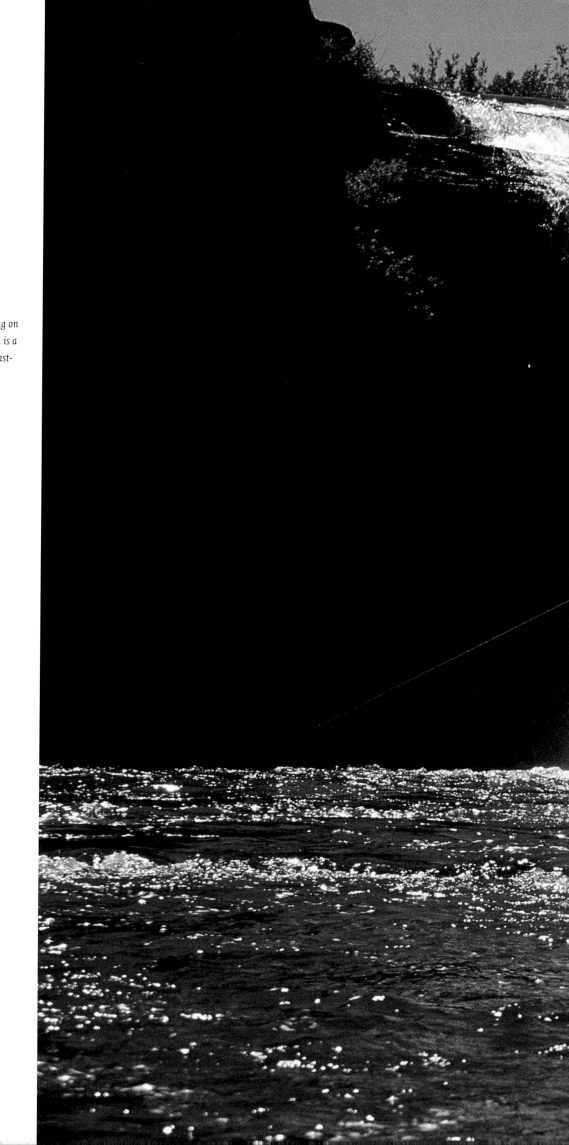

Fly-fishing using a 'fly' floating on the surface of the water as bait, is a sport practised principally in fast-flowing rivers.

The Rhine Falls at Schaffhausen near the Swiss-German border. The Rhine is unnavigable for inland shipping on this stretch of the river.

Switzerland: the Rhine Falls

The Rhine Falls lie on the River Rhine in Switzerland, close to the border with Germany. These beautiful falls lie in the upper course of the Rhine to the west of Lake Constance (Bodensee). Some 15,000 years ago, at the time of the Würm ice age, these waterfalls were created in the landscape left behind by retreating glaciers. The course of the River Rhine has been greatly influenced by glaciers. They formed an impenetrable barrier and forced the river, in the neighbourhood of the Rhine Falls, to change course to the south at various stages. Whenever the Rhine changed course, the abandoned river bed slowly filled up with sediment. The original bedrock consists of hard limestone banks that formed in the Jura Mountains. The limestones slope to the southeast and are covered with softer sediments. At the Rhine Falls, a fossil river bed of the Rhine can be found, which was eroded into the hard limestone. As a result of an advancing glacier, part of the Rhine at Neuhausen was forced to flow towards the southeast. The resulting abandoned river bed filled with sediment. The course of the Rhine then crossed over a sudden, abrupt change in the composition and in the erosion resistance of the underlying bedrock. The hard limestone is more difficult to erode than the unconsolidated sediments in the fossil river bed. A difference in height has arisen through the erosive force of the water, with the cascading water carrying off the unconsolidated sediment of the fossil river bed.

The bedrock consists of the same hard limestone over which the Rhine now pours. The difference in erosion resistance has disappeared. Percolating ground water will further undermine the rock at the top of the waterfalls, so that it will finish up collapsing and eroding the waterfall completely.

The Rhine Falls is divided into three sections: the Zürcher, the Schaffhauser and the Mühle Falls. The three waterfalls are split up by islets of jagged limestone that defy the tumbling water. They are a remnant of the original precipice over which the water used to plunge. At present the Rhine Falls

are 25 metres high and 150 metres wide, with a discharge amounting to 750,000 litres of water per second flowing over the edge, barely a fifth of that of Niagara Falls. Nonetheless, these are the largest falls in Western Europe and draw 50,000 visitors every day in the high season.

Despite the small drop these Falls still make an impression on the speed with which the water flows over the shallow lip.

The Rhine Falls serve an important old trading route: the course of the Rhine from Lake Constance to Basel. This route was mainly used from the 13th century to the middle of the 19th century to transport salt.

◁◁△ The Slettningsegg Falls in Norway. Falling and flowing water has an internal energy that is called kinetic energy. By using the water to drive a turbine, this energy can be transformed into mechanical energy, which in turn is used to power a electricity generator. In Norway 99.9% of the energy requirement is met in this way. The energy that could be produced world-wide in this manner, is, in theory, in excess of the energy requirements.

◁ Waterfalls like these could also be used to generate hydro-electric power.

Hydro-electricity

Every river with flowing or cascading water can be used to generate energy. This form of energy has been given the nickname 'white coal'. Dams are often built in the course of a river so that the water will gather behind in reservoirs. This makes the waterflow more manageable and can provide a constant year-round supply of electricity, or if desired a higher demand can be met for part of the year, by using the reservoir of water. The water is led from the reservoir through large pipes (penstocks) down into the turbines. These are situated at the foot of the dam and transform the kinetic (movement) energy into mechanical energy. This mechanical energy is then used to generate electricity. The creation of dams to provide reservoirs for the generation of electricity has serious consequences for the environment of the river valley, both above and below the dam, and is, moreover, very costly. For this reason, smaller scale projects are often preferred nowadays, often using turbines set directly into the river.

By using more rivers, including the smaller ones, to generate electricity, the increasing demand for energy can be met in an environmentally-friendly way. The country with the highest energy requirements, the United States, could, in theory, meet its entire energy requirements through large-scale use of water power. Hydrologists have calculated that the rivers in the United States transport 550 million litres of water per second over a vertical distance of 500 metres. The energy that could be generated if this was harnessed is more than sufficient to cover the entire present consumption. This is, of course, unobtainable in the short term, but it does indicate that hydro-electricity can provide a substantial proportion of the total energy requirement in the future.

Hydro-electricity is already generated at Niagara Falls. On both the Canadian and the American sides, two large tunnels have been constructed through which an average of 4.5 million litres of water per second drops down, powering the turbines below. The turbines transform the force of the water into 2,400 megawatts of electricity, enough for almost two million homes. Just how much energy, flowing water can generate is also apparent from the capacity of the La Grande complex in Canada. Here, five large hydro-electric power stations will, in the year 2000, provide ten million homes with electricity. The extent to which water power can be used to generate electricity depends on a number of factors. The relief of the land plays an important role. A country like Switzerland with almost only mountainous land has far more opportunities to use water power than a low-lying country like the Netherlands. In addition, the construction of hydro-electric power stations requires massive investment, so that only richer countries can afford to build them. If there is money available, then political choices also play a role in whether the money is used for this kind of project. World-wide, only 2% of the electricity generated comes from hydro-electric schemes. In some countries, however, water power plays a very important role in the electricity supply. The best known example is Norway, where 99.9% of the electricity used is generated using water power. Canada also meets some 70% of its electricity needs through hydro-electric generation. Other countries do not even achieve 50%, although countries such as Australia, Argentina and France do manage to generate some 25% of their electricity requirements using water power.

Small changes in the depth of a river can create rapids in the blink of an eye. The 'veil' effect shown in the photograph above was achieved by using a slower shutter time on the camera.